I0617410

ABC

Rocks and Minerals Book

AO PRESS

Jessica Lee Anderson

Copyright © 2024 by Jessica Lee Anderson
All rights reserved.

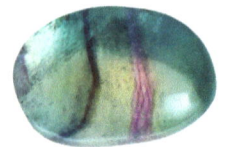

No portion of this book may be reproduced or utilized in any form or by any means – electronic or mechanical, including photocopying or recording, or by any information storage and retrieval system – without written permission from the author. For more information, visit www.jessicaleeanderson.com.

Paperback ISBN: 979-8-9899560-3-6

To the Austin Gem & Mineral Society—thanks for living out your motto of "Each One - Teach One" so beautifully. - JLA

Note that various vocabulary words have been capitalized for emphasis.

Photo credits—Front Cover: Aeya, vvoevale, Johannes.k, Genneristock; Back Cover: Vladk213 (Quartz Geode); Cover Page: Aeya, vvoevale, Johannes.k, Levon Avagyan, Gina Chiriac, Tiero (Diamond, Yellow Diamond, Blue Sapphire, Pink Sapphire); Copyright Page: vvoevale (Fluorite) KrimKate (L to R: Adventurine, Amazonite, Sodalite); Dedication Page: Kathy Feeney; p. 4: stux, Shatenka07, p. 5: Jonnysek, p. 6: vvoevale, KrimKate; p. 7: Aeya, Stellar-Serbia, p.8: KrimKate, Aeya, Aksidesign; p. 9: Difydave; p. 10: Jaanalisette; p. 11: Sarah Salahuddin, vvoevale, KrimKate; p. 12: Snapwire from Pexels, Furtseff; p. 13: vvoevale, lermannika, vvoevale; p. 14: KrimKate; p. 15: KrimKate; p. 16: vvoevale; p. 17: KrimKate; p. 18: Stellar-Serbia; p. 19: Aeya; p. 20: Nastya22, BrankoBG, WillScape; p. 21: Aeya, Nawakorn Yanthawai, sennoynoy; p. 22: Aeya; p. 23: Michel Viard; p. 24: Verbaska_Studio; p. 25: KrimKate; p. 26: SunChan, SageElyse; p. 27: KrimKate; p. 28: KrimKae; p. 29: Cagla Kohersli; p. 30: WojciechMT, malachit-obchod, LorenzoT81, Charlette Salcedo (all Agates); p. 31: Africa Images; p. 32: Michael Anderson, Scott Orr (Desert Rose Selenite), beachboy (Labradorite)

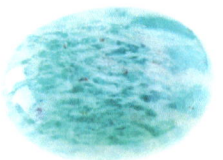

This Book Belongs to:

_____

is for Amethyst

Amethyst ranges from lilac to a deep purple, and it has been popular throughout history. It is a purple variety of a mineral called Quartz.

A a

4

is for Bronzite

Bronzite is a mineral with a bronze color that can be polished for jewelry. It has a metallic luster, meaning it shines by reflecting light.

C is for Carnelian

Carnelian ranges from a fiery orange to red to brownish red. It is a member of the Quartz mineral family.

is for Diamond

Diamonds are one of the world's hardest natural substances! Diamonds can look rough before a lapidary (a gem cutter) will create facets, or polished surfaces that show off the beauty.

E is for Emerald

Emeralds are precious gemstones along with Diamonds, Rubies, and Sapphires. Emeralds have been mined for thousands of years!

E e

F is for Fool's Gold

Fool's Gold is also known as Pyrite. Fool's Gold got its name because it has fooled many people into thinking that it is real Gold, though it has differences like how it ranks on the Mohs Hardness Scale (a scale that helps to identify minerals).

F f

 is for Gold

Gold is an element (meaning it can't be broken down through a chemical reaction) as well as a mineral. Gold is also a metal that can be shaped—it is softer compared to Fool's Gold.

G g

H is for Heliotrope

Heliotrope is also known as Bloodstone because of the red flecks inside of this gemstone. It can be found all over the world.

 is for Icicle

Icicles (and solid ice in general) have a natural crystal structure. Along with its chemistry, this makes ice a mineral, but the same is not true when it melts.

Ii

J is for Jasper

Jasper is a type of Quartz formed from small crystals. It comes in a variety of colors and patterns!

 **is for Kimberlite**

Kimberlite is a type of igneous rock formed when magma (underground molten rock) cools down and becomes hard. Diamonds and gems can be found in Kimberlite.

K k

L is for Lapis Lazuli

Lapis Lazuli is a type of rock formed from several different minerals including Fool's Gold. The bright blue colors have made this rock popular since ancient times.

L l

M is for Moonstone

Moonstone has a soft glow much like moonlight. Moonstone is often shaped and polished into a cabochon (or cab for short) that is used in jewelry—a cab is flat on the bottom and round on the top with no facets.

M m

N is for Nephrite

Nephrite is one of the minerals that is known as Jade (the other is called Jadeite). Nephrite has been used for weapons, ornaments, and jewelry throughout the ages.

N n

O is for Opal

Opals are iridescent—this means that their showy colors seem to change when you look at it from different angles. Some opals are natural though others are man-made.

P is for Peridot

Peridot is famous for its bright yellowish green color, and unlike other gemstones, it only comes in one color. Peridot often occurs in volcanic rocks.

Pp

Q is for Quartz

Quartz is a mineral that's found easily in all types of bedrock (the layer of hard, solid rock underneath soil and gravel). Quartz comes in many colors, patterns, and textures.

Qq

R is for Ruby

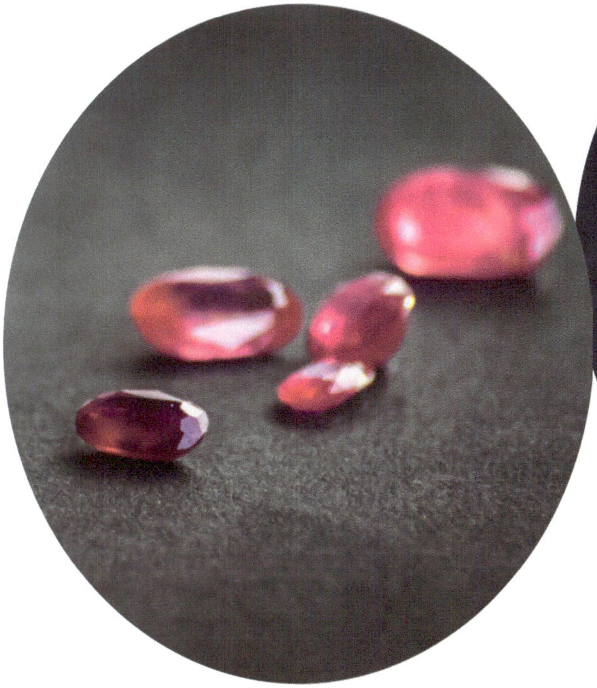

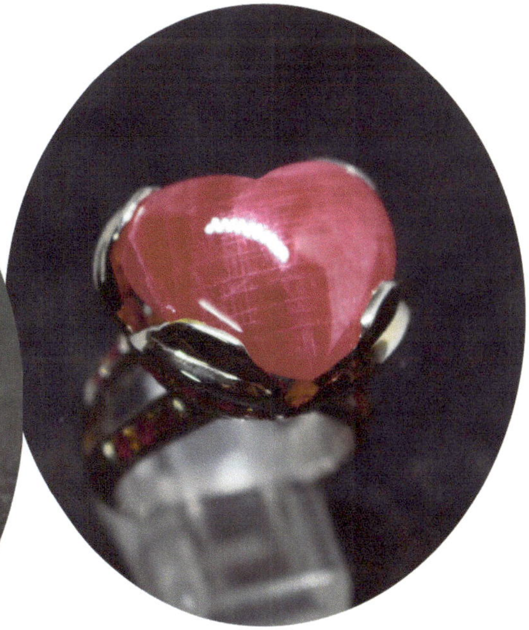

Rubies are rare, and they range from pinkish red to blood-red. They are quite strong!

Rr

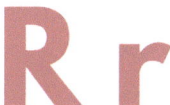

 **S** is for Sapphire

Sapphires are similar to Rubies, though they have different colors. Kings and queens have worn Sapphires throughout history.

S s

T is for Turquoise

Turquoise is a blue to green mineral that is opaque, meaning you can't see through it. Turquoise has been popular for thousands of years, and it is the only gemstone that has a color named after it.

is for Unakite

Unakite is a pink and green type of igneous rock named after the Unaka Mountain Range. Unakite is used in various building projects for things like tiles and stairs, and it is even crushed up for highway construction.

Uu

 is for Violane

Violane is also called Violan or Blue Diopside. It is rich in a chemical element named Manganese that is also important for the human body.

Vv

 **is for Watermelon Tourmaline**

Watermelon Tourmaline is an interesting variety of Tourmaline with watermelon-like colors. It is not just one mineral but a group of similar minerals.

 is for Xonotlite

Xonotlite is also known as Eakleite. This mineral can glow under ultraviolet (UV) light (meaning it is luminescent).

 **is for Yellow Citrine**

Yellow Citrine is sometimes called Lemon Citrine. Natural Citrine is a rare type of Quartz.

Z is for Zultanite

Zultanite is transparent, meaning light can pass through it. Zultanite is found in the mountains of Turkey.

Z z

5 Rock and Mineral Facts:

 Minerals are natural substances that have distinct chemical and physical properties. They do not come from animals or plants.

 Rocks are made of minerals.

 The study of rocks and minerals is called Geology. Geology has several different branches of study like Paleontology and Mineralogy.

 Mining is the process of removing minerals from Earth.

 Some of the most common mineral groups are the Silicates, Carbonates, Sulfates, Halides, and Oxides.

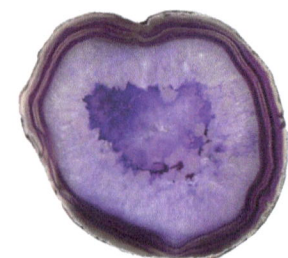

Jessica Lee Anderson is an award-winning author of over 75 books for young readers. Jessica lives near Austin, Texas with her daughter, Ava, and husband, Michael. They are all rockhounds! Desert Rose Selenite and Labradorite are some of Jessica's favorites. You can learn more about Jessica by visiting www.jessicaleeanderson.com.

Check out these other titles:

ABC
Reptile Book
Jessica Lee Anderson

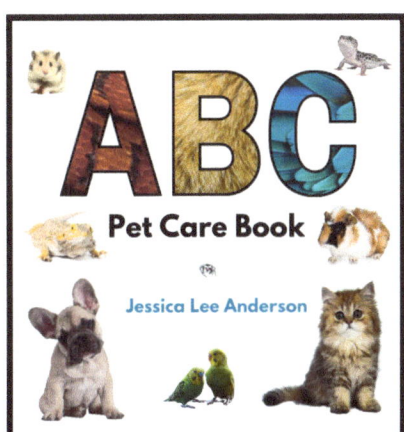

ABC
Pet Care Book
Jessica Lee Anderson

ABC
National Parks Book
Jessica Lee Anderson

www.ingramcontent.com/pod-product-compliance
Lightning Source LLC
Chambersburg PA
CBHW041620120626
46551CB00003B/512